MEMOIR

OF

JOHN MICHELL

MEMOIR

OF

JOHN MICHELL

M.A., B.D., F.R.S.

Fellow of Queens' College, Cambridge, 1749
Woodwardian Professor of Geology
in the University 1762

BY

SIR ARCHIBALD GEIKIE

O.M., K.C.B., D.C.L., D.Sc., F.R.S.

CAMBRIDGE
AT THE UNIVERSITY PRESS
1918

CAMBRIDGE
UNIVERSITY PRESS

University Printing House, Cambridge CB2 8BS, United Kingdom

Published in the United States of America by Cambridge University Press, New York

Cambridge University Press is part of the University of Cambridge.

It furthers the University's mission by disseminating knowledge in the pursuit of education, learning and research at the highest international levels of excellence.

www.cambridge.org
Information on this title: www.cambridge.org/9781107623781

First published 1918
First paperback edition 2014

A catalogue record for this publication is available from the British Library

ISBN 978-1-107-62378-1 Paperback

CONTENTS

INTRODUCTION PAGE 1

BIOGRAPHICAL SKETCH 3

SCIENTIFIC WORK

I. Contributions to Geology 24

II. Contributions to Physics 72

III. Contributions to Astronomy 91

INDEX 106

JOHN MICHELL

Among the men of science in England during the latter half of the eighteenth century there was one specially remarkable for the wide range of his genius and the originality of his methods of research. As Rector of a quiet country parish in Yorkshire, he lived remote from the centre of the intellectual life of his day, but in that retreat he had time and opportunity for reflection and experiment. Moreover, as he was able to visit London each year, he could keep in touch, not only by correspondence but by personal contact, with the leaders of enquiry. Though much esteemed and respected by his contemporaries, he has perhaps hardly received from subsequent generations the recognition to which the merit of his work justly entitles him. It is true that some historians who have recorded the progress of the sciences to which he devoted his attention have alluded more or less fully to his published papers. But it is not until a review is made of his contributions to each of the sciences of geology, physics and astronomy that an adequate conception can be formed of the place that is due to him in the history of English science.

Recent researches among the archives of the Royal Society and of its dining Club brought the name of this modest investigator so frequently before me as to rouse my interest in his career. I was induced to search for any personal details regarding him that might still be recoverable, and to peruse such of his writings as I had not previously read. As the result of this enquiry I have thought it to be my duty to bring his life and his solid achievements in science more prominently to notice. Hence the origin of the present Memoir.

JOHN MICHELL, the friend of Henry Cavendish and Joseph Priestley, has left no record of his life beyond his published writings. A few of his letters have survived. Several of them addressed to Sir William Herschel have been found among that philosopher's manuscripts, and are quoted in the collected edition of his *Papers*. A long and hitherto unpublished letter from Michell to Cavendish has been preserved among the papers of that great man, and is inserted in the present Memoir[1].

It is not quite certain where and when Michell was born. Probably his native place was Nottingham, and the year of his birth 1724. Of his parentage nothing appears to be known. The earliest accounts of him which have been recovered are preserved in the registers of Queens' College, Cambridge, where a full

[1] This letter was known to Dr George Wilson, Cavendish's biographer, and is referred to by him in the *Life* (*postea*, p. 47).

record has been kept of his College life, from the time of his admission as Pensioner on 17th June 1742, until, after taking his degrees and filling many offices during a residence of twenty-one years, he quitted Cambridge for a rectory in the country[1]. He is entered in these records as from Nottingham.

The year after his reception into Queens' he was elected Bible-Clerk and held this office for two years. Again for three years, from 1747 to 1749 he filled the same post[2]. He did not take his Bachelor's degree until 1748. His name appears as fourth wrangler in the list for 1748–9, which was the second competition after the institution of the wranglership. On 30th March 1749 he was chosen Fellow

[1] The present President of Queens' College, the Rev. T. C. Fitzpatrick, has been so good as to collect for me all the details of College life that are here given.

[2] The duty of this officer appears to have been to read the Bible in hall, for which a remuneration of one shilling a week was allowed, afterwards slightly increased.

of his College. Thereafter for some fifteen years he continued to fill various lectureships and other offices at Queens'. He was Tutor of the College from 1751 to 1763; Praelector in Arithmetic in 1751; Censor in Theology in 1752–4; Praelector in Geometry in 1753; Praelector in Greek in 1755 and 1759; Senior Bursar in 1756–8; Praelector in Hebrew in 1759 and 1762; Censor in Philosophy and Examiner in 1760. He took his degree of Master of Arts in 1752 and Bachelor of Divinity in 1761. He was nominated Rector of St Botolph's, Cambridge, on 28th March 1760, and held this living until June 1763, when he left Cambridge on being collated to a rectory in the country.

The registers of Queens' College furnish information as to the modest payments made in the eighteenth century to the officials by whom the work of the College was performed. In Michell's case we learn that the largest sum paid to him as

Bible-Clerk was £5. 3*s.* 10*d.* for the year 1748. Again in 1753 his stipend as Fellow amounted to £9, that of his theological Censorship to £8, and that of his examinership to £2.

Besides these College duties he held from time to time some University appointments. In July 1754 he was elected to the office of Moderator for the following year. In 1755 he was appointed Taxator[1] and on 24th June 1762 Scrutator for the following year. But the most important office conferred upon him was the Woodwardian Professorship of Geology, to which, near the end of the year 1762, he was appointed by Colonel King, the last surviving executor of John Woodward

[1] The Master of St John's informs me that "the Taxators were appointed by the Colleges according to the cycle for Proctors. They regulated the markets, examined the assize of bread, the lawfulness of weights and measures, and called all abuses and defects into the court of the Commissary. The Scrutators seem to have been assistants to the Proctors. The Proctors read the Graces and took the votes in the Regent House; the Scrutators did the like in the Non-Regent House."

who by his will dated in 1727 founded the Chair. Michell did not hold the office for quite two years, having to vacate it on his marriage in 1764. There appears to be no evidence that during his short tenure of the office he ever gave geological lectures[1]. But the intimate acquaintance with geological phenomena shown in his essay on Earthquakes, communicated to the Royal Society in the spring of 1760, proves that he was well qualified to lecture on a subject which he had pursued with zeal in the field. It is difficult to believe that he did not impart to his undergraduate friends some of the knowledge which he had gained in many traverses across the southern counties, if indeed he did not take them with him in some of his rambles.

A brief description of Michell's personal appearance in his College days, penned by a contemporary diarist and preserved among the manuscripts of the

[1] *Life of Adam Sedgwick*, vol. I, p. 192.

British Museum, may be quoted here. "John Michell, BD is a little short Man, of a black Complexion, and fat ; but having no Acquaintance with him, can say little of him. I think he had the Care of St Botolph's Church, while he continued Fellow of Queens' College, where he was esteemed a very ingenious Man, and an excellent Philosopher. He has published some Things in that way, on the Magnet and Electricity[1]."

Although his time was evidently much engaged in the various official duties that devolved upon him in Cambridge, there is proof that he had already launched upon his career of physical research and experimentation within the walls of his College. In 1750, the year after he obtained his Fellowship and when he was some six-and-twenty years of age, he published at Cambridge a little volume on Artificial Magnets to which further reference will

[1] Cole MSS. XXXIII, 156 (Add. MSS. Mus. Brit., 5834).

be made on a later page. As he was fond of constructing his own apparatus, his rooms at Queens' with all his implements and machinery would sometimes wear the look of a workshop, and were no doubt the theme of much amused wonderment among both Fellows and undergraduates.

But these mechanical operations and experiments indoors were only a part of the scientific occupations with which he employed his leisure. As above stated, there can be no doubt that he was in the habit of making what would now be called geological excursions, in which he interested himself in noting the distribution and sequence of the rock-formations in the southern counties of England. It was only by such practical field-work that he could gain the remarkably accurate conception of the structure of the stratified portion of the earth's crust embodied in his Earthquake paper of 1760. This epoch-making essay was read to the Royal Society in sections at five successive evening meetings. The active College Preceptor and Bursar was now introduced into the centre of the scientific life of the time, where he was warmly welcomed. Immediately after the reading of the first portion of his paper some of the Fellows of the Society drew up and signed a certificate in favour of his election into the Society. Within a fortnight, and before the reading of his paper was ended, the certificate was in the hands of the Council. It ran as follows:

> The Rev. Mr John Michell M.A. Fellow of Queens' College, Cambridge, who has recommended himself to the Publick by his Experiments in Magnetism, and has lately communicated to this Society a Dissertation upon Earthquakes, being very desirous of the honour of becoming a member of the Royal Society, We, whose names are underwritten, recommend him as a gentleman extremely well qualify'd for that honour.
>
> London, *March* 6, 1760

The first names on the list of signatures are those of the Secretary of the

Society, Dr Thomas Birch, and of Michell's contemporary at Queens' College, the active and broad-minded Sir George Savile, Bart., who, after serving in his youth against the Jacobite rebels in 1745, spent a busy and useful life in Parliament as member for Yorkshire. Next comes the name of Dr Gowin Knight, Copley Medallist, and first Principal Librarian of the British Museum, whose researches in magnetism would especially draw him towards Michell. The other signatures include those of Dr Matthew Maty, afterwards Secretary of the Royal Society and Principal Librarian of the British Museum; Daniel Wray, the antiquary; and John Hadley, another of the Fellows of Queens' College, who a few years before had been appointed Professor of Chemistry at Cambridge, and with whom Michell would doubtless have much discussion of scientific matters. Michell was duly elected a Fellow of the Royal Society on 12th June 1760. It is worthy of notice that the immediately preceding election was that of Henry Cavendish, and that the two names stand together in chronological sequence in the *Record* of the Society.

In the spring of the year 1763 Michell gave up his residence at Cambridge and became Rector of Compton in the valley of the Itchin, which winds from Winchester to the sea. It was not improbably with a view to his marriage that this change of abode was made. The following announcement appeared in the *Cambridge Chronicle* of 8th September 1764. "A few days ago the Rev. Mr Michell, Rector of Compton, near Winchester, late Fellow of Queens' College, Cambridge, and Woodwardian Professor of Fossils in this University, was married to Miss Williamson, a young lady of considerable fortune, near Newark in Nottinghamshire." He did not long retain his living at Compton, for on 23rd January 1765, he was collated

Rector of Havant, Hants. The prospect of a happy married life at this home was rudely dissipated in the autumn of that same year, when his young wife died at Newark[1]. Two years later, on 3rd October 1767, he was instituted Rector of Thornhill, near Dewsbury in Yorkshire, and remained in this office as long as he lived. He subsequently married again[2]. He appears to have had only one child, a daughter, probably by his first wife. This daughter married and is said to have died at an advanced age somewhere about the year 1840.

In the summer of 1871 there appeared in the journal called the *English Mechanic*[3] a communication which gave some previously unpublished information about John Michell. As the writer of the letter stated that he was the great-grandson of

[1] *Cambridge Chronicle*, 12th October 1765.

[2] In the church at Thornhill, the burial register records that Ann Michell, relict of the late Rev. J. Michell, Rector, died 6th November 1818.

[3] Vol. XIII, p. 310 (16th June 1871).

the philosopher, and that he derived his information from his grandmother, Michell's daughter, considerable importance has naturally been attached to his contribution, and its statements have been quoted, though sometimes with hesitation, in various biographical notices of his eminent ancestor. It ran as follows:

After William Herschel's appointment as organist of Halifax, he became acquainted with the Rector of Thornhill, (a village about nine miles from Halifax and six from Wakefield) whose name was John Michell—a man of fortune, whose whole life was devoted to science, and whose writings are to be found in considerable numbers in the journals of the Royal Society, during the latter half of the last century, one of the most prominent papers being that on the great Lisbon Earthquake of 1755. John Michell may perhaps be better known as the builder of the mathematical bridge across the Cam at Queens' College, Cambridge[1]. He was no mean violinist in

[1] This family tradition is probably an exaggeration of any connection which Michell may have had with the bridge. He obtained his Fellowship at Queens' on 30th March 1749. In October of that year it is recorded in the *Magnum Journale* of

his day, and his soirées—where not only the first musical talent, but also the first scientific men of the day, such as Cavendish, Black and Priestley used to meet occasionally—were well known in the West Riding of Yorkshire, and to which Wm Herschel used to come to perform on the violin. At the period of these visits Michell was and had long been engaged in making what was at that time a large telescope—a ten-foot reflector. The perfect combination for a perfect reflector, and the grinding the same, had long occupied Michell's attention, in which he at last succeeded, and I believe I am correct in saying that Herschel there became a willing and able pupil, and obtained the germs of his great astronomical renown. At the death of John Michell, all his scientific apparatus were sent to Queens' College, Cambridge, save and except his large reflecting telescope, which by purchase or gift came into the possession of Wm Herschel. I have been told by the only

the College that Mr Etheridge was paid £21 for the design and model of the bridge. The construction was not completed until September of the following year when the cook was paid 17*s*. 9*d*. for a supper to the workmen on finishing their work. Michell could not fail to be interested in the operations, and may quite possibly have been able to give useful help to the designer, as well as to James Essex, the builder of the bridge.

child of Michell, who died about thirty-five years ago, at the age of upwards of eighty, and was intimate with Herschel, that he told her that the principal part of his observations had been made with her father's telescope, which he found more convenient than his own larger one. The Rev. John Michell, I have also been informed, was the inventor of an apparatus for ascertaining the weight of the world, which is known as that of Cavendish.... I am the grandson of Michell's only daughter, from whom I heard much, and I was also a pupil, more than fifty years ago, of an old clergyman, who had in early life been for several years the curate of Thornhill under Michell. [Signed] Khoda Bux.

That the memory of the daughter or that of the great-grandson, or of both together, had failed can easily be proved. Some of the statements in the communication are curiously inaccurate. William Herschel undoubtedly passed some of his early years in Yorkshire where he played the hautboy in the band of the Durham militia and performed on the violin at public concerts in Wakefield and Halifax. The family tradition that Michell was "no mean violinist" and that he gave musical parties is not improbably true. But it appears to be quite certain that he and Herschel did not make acquaintance with each other in Yorkshire, and that the Rector of Thornhill did not instruct the future illustrious astronomer in the art of grinding specula. Herschel was appointed organist at Bath in the autumn of 1766 and removed to that city at the beginning of the following year, before Michell succeeded to the living at Thornhill. There was thus no opportunity for their meeting in Yorkshire. Herschel at Bath was absorbed in his duties as a professional musician and did not enter upon the study of astronomy till 1773, and next year began to grind specula. The two men had never met nor exchanged letters up to the spring of the year 1781[1].

The legend that the discoverer of the planet Uranus received his first lessons in

[1] *Scientific Papers of Sir William Herschel* (1912), vol. I, p. xxxii.

telescope construction from the learned Rector of Thornhill must thus be dismissed as unfounded. Each of these two men of science worked independently and ground his own mirrors. They were first brought into correspondence with each other through the medium of Dr Wm Watson, F.R.S., who, being at Bath and knowing Herschel there, sent to Michell an account of what the astronomer had been doing. On 21st January 1781, Michell replied: "I look upon myself as very much obliged to you for your favour from Bath, and particularly for the very interesting account, both of what Mr Herschel has done and what he has seen, both of which seem to be very important. I shall be very happy if I should be able to succeed as well, or near as well, as from your account he seems to have done, and I should be very glad of the favour of his correspondence: at the same time I think it very probable that I may be more likely to learn from him what may be useful to myself, than he to learn anything from me[1]." A brief correspondence followed between the two astronomers, but they probably met occasionally at the Royal Society's rooms, the table of the Royal Society Club, or other meeting places in London.

The search for further personal records of the Rector of Thornhill has not been successful. His life in the Yorkshire parish was doubtless quiet and uneventful, full of pastoral duty, well discharged, but with leisure for the prosecution of scientific work. This work was undertaken for its own sake, and much of it was probably never published, though it was

[1] *Scientific Papers of Sir William Herschel*, vol. 1, pp. xxxi, xxxii. The two astronomers never met nor exchanged letters before 12th April 1781, on which date Michell addressed a letter to Herschel dealing with mirrors, and the relative merits of different types for large and small apertures. *Op. cit.* p. xxxii. It is interesting to know, from the testimony of Herschel himself, that after Michell's death, he purchased his large telescope. See *postea*, p. 96.

always at the service of any fellow-worker interested in the same subjects. Though living so far from London, Michell was nevertheless able to maintain intimate relationship with some of the most eminent men of science of his time. He appears to have had some appointment or duty which for a number of years took him annually to London, and gave him the opportunity of attending the meetings of the Royal Society and cultivating the companionship of his scientific friends. From the year 1758 onwards till towards the close of his life he continued to be a frequent guest at the weekly dinners of the Royal Society Club. There would seem, indeed, to have been a kind of friendly rivalry among the members of the Club in securing him as a guest at these dinners, and in this social competition his friend Henry Cavendish took part. Thus during the summer of 1784 he dined seven times with the Club. Two years later, when he spent the months of May and June in the Capital, he was a guest every week during his stay, and similar hospitality awaited him at the "Crown and Anchor" tavern as long as he lived[1].

During his journeys from and to his Yorkshire home, now by one route, now by another, on horseback or by carriage, or in such public coaches as then plied on the roads between London and the Midlands, Michell appears to have paid close attention to the outcrops and succession of the rock-formations across which he travelled. There was probably no one else so familiar as he with the various quarries, pits and other exposures of these rocks to be seen from the high roads. In the leisurely fashion of those days he would sometimes halt for a night or two on the way, and on these occasions would take advantage of the opportunity of obtaining the evidence to be found in

[1] *Annals of the Royal Society Club*, pp. 74, 77, 165.

any fresh opening of the ground. By these means and by special excursions for the purpose of geological investigation which as we have seen he began to make in his College days, he acquired a broader and more accurate conception of the geological structure of the southern half of England than any of his predecessors. Moreover, his rectory lay in the heart of the great Yorkshire coal-field where the progress of the mining industry continually brought geological questions to his notice.

It was at Thornhill, during the leisure which his clerical duties permitted, that he was able to carry on those important investigations in physics and astronomy with which his name will always be associated. In the same quiet retirement he devised and constructed the various forms of original and ingenious apparatus, with which he solved or illustrated the problems that presented themselves to his ever active intellect. He continued to reside at Thornhill till his death on 21st April 1793 in the sixty-ninth year of his age[1].

In reviewing the scientific work accomplished by John Michell it will be convenient to consider it in three sections: first, his contributions to Geology; second, his contributions to Physics; and third, his contributions to Astronomy.

[1] Michell's tombstone at Thornhill records in characteristic eighteenth century language his life, work and character: "In the Chancel of this Church are deposited the Remains of the Revd. Jn. Michell, B.D., F.R.S., and 26 years Rector of this Parish. Eminently distinguished as the Philosopher and the Scholar, he had a just claim to the character of the real Christian, in the relative and social duties of life. The tender Husband, the indulgent Father, the affectionate Brother and the sincere Friend were prominent features in a character uniformly amiable. His Charities were not those of Ostentation, but of feeling; His strict discharge of his Professional duties, that of principle, not form. As he lived in possession of the esteem of his Parishioners, so he has carried with him to the grave their regret. He died the 21st April 1793, in the 69th year of his age."

I

CONTRIBUTIONS TO GEOLOGY

THE subject of the structure of the earth's crust appears to have engaged Michell's attention long before he left Cambridge, though in this as in other investigations, he was in no hurry to publish his observations. He was ultimately led to embody the results of his studies in this subject in the paper on Earthquakes which he communicated to the Royal Society in the spring of the year 1760[1]. This remarkable Essay is not only a dissertation on earthquakes, but contains an exposition of the structure of the terrestrial crust

[1] The full title of this paper was as follows: "Conjectures concerning the Cause, and Observations upon the Phaenomena of Earthquakes; particularly of that great Earthquake of the first of November 1755 which proved so fatal to the City of Lisbon, and whose Effects were felt as far as Africa, and more or less throughout almost all Europe; By the Reverend John Michell M.A. Fellow of Queens' College, Cambridge." It appeared in vol. 51 (1760) of the *Philosophical Transactions*, pp. 566–634.

which he had worked out in the course of his journeys and excursions from Cambridge over the southern counties of England. As this stratigraphical work was continued through the rest of his life I shall consider it in detail after first dealing with the portion of the paper more immediately concerned with the cause and phenomena of earthquakes.

1. *The Cause and Phenomena of Earthquakes*

The attention of the civilised world was strongly drawn to the long series of earthquakes which culminated in the appalling catastrophe that overwhelmed Lisbon and shook the greater part of Europe on 1st November 1755. Many accounts of the facts as well as attempted explanations of them were printed in the current literature of the time. In particular, the Royal Society devoted the forty-ninth volume of the *Philosophical Transactions* to a large collection of reports on the subject. By far the most important contribution to the scientific discussion of the nature and cause of these earth-tremors was this paper by Michell. It contained the first attempt to suggest a natural cause of earthquakes, and to explain the nature of seismic movements.

In attacking the problem Michell started with the fundamental postulate of the existence of "subterranean fires." It was the general belief of the day, which he shared, that these "fires" resulted from the spontaneous ignition of pyritous inflammable strata, such as coals and carbonaceous shales lying at variable distances beneath the surface of the earth. Werner and his followers were so convinced that such was the origin of these "fires," to which they attributed the existence of volcanoes, that they believed volcanic action to be of comparatively late date in the history of the globe, seeing that it could not arise till vegetable growths had long flourished on the earth and been buried under sediments to become beds of coal. Michell argued that if a large body of water should be let down suddenly upon one of these fires a vapour would be produced, the quantity and elastic force of which might be fully sufficient to account for the origin of earthquakes. He pointed out the frequency of these concussions in volcanic districts and, like the Wernerian geognosts, he connected the phenomena of volcanoes with the same subterranean fires, the vents and craters marking the position of weaker portions of the outer shell of the earth across which the vapour generated below could force its way to the surface, often carrying with it an outpouring of molten material. The energy with which this escape was effected appeared to him to be evident from the vast size of the blocks of rock projected during volcanic eruptions and the great distances to which such masses of stone were often thrown. "If," he asked, "when the vapours find a vent, they are capable of shaking the country to the distance of ten or twenty miles round, what may we not expect from them when they are confined?"

He stated that "the motion of the earth in earthquakes is partly tremulous, and partly propagated by waves which succeed one another, sometimes at larger, sometimes at smaller distances, and this latter motion is generally propagated much further than the former[1]." He believed that both of these motions could be accounted for by the steam generated as assumed. "Let us suppose," he remarked, "the roof over some subterraneous fire to fall in. The earth, stones, etc., of which it was composed would immediately sink in the melted matter of the fire below: hence all the

[1] Art. 11 of the paper. Michell appears to have been the first to point out that the earthquake travels in successive waves through the earth.

water contained in the fissures and cavities of the part falling in would come in contact with the fire and be almost instantly raised into vapour. From the first effort of this vapour, a cavity would be formed (between the melted matter and superincumbent earth) filled with vapour only, before any motion would be perceived at the surface of the earth: this must necessarily happen, on account of the compressibility of all kinds of earth and stones, etc.; but as the compression of the materials immediately over the cavity would be more than sufficient to make them bear the weight of the superincumbent matter, this compression must be propagated on account of the elasticity of the earth in the same manner as a pulse is propagated through the air. And again, the materials immediately over the cavity, restoring themselves beyond their natural bounds, a dilatation will succeed to the compression; and these two following each other alternately, for some time, a vibratory motion will be produced at the surface of the earth. If these alternate dilatations and compressions should succeed one another at very small intervals, they would excite a like motion in the air and thereby occasion a considerable noise. The noise that is usually observed to precede or accompany earthquakes, is probably owing partly to this cause, and partly to the grating of the parts of the earth together, occasioned by that wave-like motion before mentioned" (Art. 56).

"As a small quantity of vapour almost instantly generated at some considerable depth below the surface will produce a vibratory motion, so a very large quantity (whether it be generated almost instantly, or in any small portion of time) will produce a wave-like motion. The manner in which this wave-like motion will be propagated may, in some measure, be represented by the following experiment. Suppose a large cloth or carpet (spread upon a floor) to be raised at one edge, and then suddenly brought down again to the floor; the air under it, being by this means propelled, will pass along, till it escapes at the opposite side, raising the cloth in a wave all the way as it goes. In like manner, a large quantity of vapour may be conceived to raise the earth in a wave, as it passes along between the strata, which it may easily separate in an horizontal direction, there being, as I have said before, little or no cohesion between one stratum and another. The part of the earth that is first raised, being bent from its natural form, will endeavour to restore itself by its elasticity, and the parts next to it, beginning to have their weight supported by the vapour, which will insinuate itself under them, will be raised in their turn, till it either finds some vent, or is again condensed by the cold into water, and by that means prevented from going any further" (Art. 58).

It is thus evident that Michell conceived the vapour to continue to force its way onward between the strata as far, at least, as the earthquake continued to be felt at the surface. His familiarity with the regular and gently inclined stratification of the rocks in the southern part of England, and his experience that at the surface of the ground they may for the most part be easily split open along their bedding-planes led him to this belief. He imagined that the vapour might be propelled even to the extreme limits of the area affected by an earthquake, for he remarks in one place that "the shortest way that the vapour could pass from Lisbon to Loch Ness was under the Ocean" (Art. 98 *note*).

The same ingenious paper showed how the centre or focus from which an earthquake is propagated may be ascertained first, from observation of the different directions from which the shock arrives at several distant places: "if lines be drawn in these directions the place of their common intersection must be nearly the place sought"; secondly, from the time of arrival of the earthquake at different places; and thirdly, from the successive arrivals of the great sea-wave. The greatest degree of exactness is obtainable "in those cases where earthquakes have their source from under the ocean, for the proportional distance of different places from that source may be very nearly ascertained by the interval between the earthquake and the succeeding wave: and this is the more to be depended on, as people are much less likely to be mistaken in determining the time between two events, which follow each other at a small interval, than in observing the precise time of the happening of some single event" (Arts. 90–93).

By way of example, the author, making use of his three indications of origin, computed that the focus of the Lisbon earthquake lay under the Atlantic Ocean, at the distance of a degree of a great circle from Lisbon, and a degree and a half from Oporto.

With singular prescience he believed that probably earthquakes are not of comparatively deep-seated origin—a conclusion which is now generally held by seismologists. He was well aware that data were then wanting on which to base any satisfactory estimate as to the depth of the focus, which might vary considerably, but he thought that "some kind of guess might be formed concerning it." Hazarding what he called "a random guess," he believed that the depth at which the impulse of the Lisbon earthquake started, "could not be much less than a mile or a mile and a half, and probably did not exceed three miles."

In considering the merit of Michell's Earthquake paper as a contribution to science we must remember that he himself modestly offered it as pretending to be no more than "conjectures." That so many of his "conjectures" and "guesses" have been confirmed in later stages of the progress of science is a striking proof of his rare genius. He accepted the current doctrine of the day regarding the existence and nature of "subterraneous fires." That doctrine has long been exploded. The general high temperature of the deeper parts of the earth's crust has been proved by abundant evidence, though the problem of the cause of this internal heat cannot yet be regarded as solved.

It seems not improbable that some of the earthquakes in volcanic regions may be produced, as Michell suggested, by the sudden descent of quantities of surface water upon the molten magma below. Extinct craters are apt to become lakes or tarns by gathering into their basins the results of the atmospheric precipitation of moisture. The sudden collapse of the floor of one of these craters, filled in this way, and the descent of a large body of water upon molten lava below would conceivably give rise to a minor earthquake. Michell therefore suggested what may be a *vera causa* in volcanic areas.

Although his conjecture that the travelling of the earthquake waves along the surface of the earth is due to the propulsion of concomitant waves of vapour between the strata underneath, has never been accepted, we should remember that it included the first recognition or at least premonitory suggestion of the potency of highly heated aqueous vapour as a subterranean dynamical agency in geological operations. Whether, if suddenly generated at a local centre, in the manner he supposed, superheated steam could force its way between the strata for some distance, as he imagined, may be doubted. But since his time, and in a manner never dreamt of by Michell, aqueous vapour has been ascertained to play a stupendous part in volcanic activity. We now know that the internal magma which underlies volcanic regions contains a vast volume of aqueous and other vapours which are held absorbed or dissolved in its molten mass under the immense pressure that prevails deep below the surface. When this pressure is lessened as the lava mounts in the throat of a vent, the imprisoned vapours escape with explosive violence, blowing the liquid material into the finest dust, which may fill the air and darken the sky over a wide stretch of country, while at the same time the still steaming lava may issue in copious streams from the crater. Moreover, geologists have long been familiar with the fact that besides being erupted at the surface by the enormous expansive force of these absorbed vapours, the magma has in many places forced its way horizontally through the terrestrial crust, often between the planes of stratification, over wide areas and for long distances. The Intrusive Sheets or Sills, so abundant in the British Isles, dating from Archaean time up to that of the Tertiary basalt-plateaux, are stupendous monuments of the part which the propulsive force of the vapours in the subterranean magma has played in the past history of the globe. The most noted of them, the great Whin-Sill of the north of England, averages from 80 to 100 feet in thickness and has been injected, possibly at successive intervals, between and across the Carboniferous strata, over an area of probably not less than 1000 square miles. With what type of earthquake the extravasation of such great sheets of molten material would be accompanied may be left to the imagination.

It is remarkable that John Michell, who recognised the influence of elastic compression in generating the vibratory movement in an earthquake, should not have advanced still farther, and have perceived that the explosive shock to which he attributed the earthquake must of itself give rise to a wave of elastic compression in the earth, which starting out in all directions may travel thousands of miles before becoming so feeble as to be no longer sensible to ordinary observation.

II. *The Structure of the Stratified Part of the Earth's Crust*

The most important part of Michell's Earthquake paper, considered as a landmark in the history of geological science, is the account which it contains of what is now known as "the crust of the Earth." Earlier in the century John Strachey[1] had shown that in the south-west of England the materials of the visible part of the earth had not been promiscuously accumulated, but had been laid down in a recognisable succession which he traced in due order from the Coal up to the Chalk. He further perceived that while the coal strata are all more or less inclined, their upturned edges are overlain by the Red Marl and later formations in horizontal

[1] *Phil. Trans.* vol. 30 (1719), p. 968; vol. 31 (1725), p. 395.

beds. Michell's observations, which were all made previous to 1760, when his paper appeared, carried the subject considerably further. It must be remembered that the importance of organic remains in stratigraphy was still unknown. Indeed, there were probably naturalists still surviving who, if they did not regard these remains as "sports of nature," firmly believed them to be memorials of Noah's Flood, during which the whole vast thickness of stratified formations was supposed to have been deposited. Even such a shrewd observer as John Woodward, founder of the Chair of Geology at Cambridge, entertained this belief, and thought that the fossils had arranged themselves according to their weight, the heavier shells and bones sinking into the deeper parts of the sediments in the diluvial waters, while the lighter organisms settled down in the upper layers.

Over the whole of the region in the south of England which he was able personally to examine Michell found that the rocks everywhere displayed a striking order and regularity. The earth, he wrote, "is not composed of heaps of matter casually thrown together, but of regular and uniform strata. These strata, though they frequently do not exceed a few feet, or perhaps a few inches, in thickness, yet often extend in length and breadth for many miles, and this without varying their thickness considerably. The same stratum also preserves a uniform character throughout, though the strata immediately next to each other are very often totally different. Thus, for instance, we shall have, perhaps, a stratum of potter's clay; above that a stratum of coal; then another stratum of some kind of clay; next a sharp grit-sandstone; then clay again; and coal again above that; and it frequently happens that none of these exceeds a few yards in thickness. There are, however, many instances, in which the same kind of matter is extended to the depth of some hundreds of yards; but in all these, a very few only excepted, the whole of each is not one continued mass, but is again subdivided into a great number of thin laminae, that seldom are more than one, two or three feet thick, and frequently not so much" (Art. 38).

This careful observer next describes with minute precision the system of perpendicular fissures (or what are known as *joints*) by which the stratified rocks are so abundantly traversed. He notices the frequent bent position of the strata, and shows that their inclination increases as they are traced towards the mountains, which are generally, if not always, formed out of the lower, and therefore older rocks. He illustrates the subject by the following experiment. "Let a number of leaves of paper, of several different sorts or colours, be pasted upon one another: then bending them up together into a ridge in the middle, conceive them to be reduced again to a level surface by a plane so passing through them, as to cut off all the part that had been raised ; let the middle now be again raised a little, and this will be a good general representation of most, if not of all mountainous countries together with the parts adjacent, throughout the whole world" (Art. 43).

This simple but ingeniously contrived model indicates how clearly its author had grasped some of the main facts which modern geology has brought to light. Thus he recognised that the sequence of stratified formations has occasionally been interrupted by upheavals whereby, along certain lines of elevation, these formations have been exposed to the action of the various denuding forces of nature by which, if the denudation continued long enough, the upraised tract would be reduced approximately to a plane, on which any subsequent deposits would, in modern phrase, lie unconformably.

From this arrangement of the stratified part of the earth's crust, as he points out,

“ we ought to meet with the same kinds of earths, stones and minerals, appearing at the surface in long narrow slips, and lying parallel to the greatest rise of any long ridges of mountains ; and so in fact we find them.” He remarks that in Great Britain, the main trend of the outcrops runs nearly north by east and south by west. He notes also that in the course of the earth-movements to which the terrestrial crust is subject, the strata have not escaped rupture. “ The whole set of strata on one side of a cleft are sunk down below the level of the corresponding strata on the other side,” and he sagaciously adds that, “ if in some cases this difference in the level of the strata on the different sides of the cleft should be very considerable, it may have a great effect in producing some of the singularities of particular earthquakes” (Arts. 44, 50).

It should be remembered that all the geological observations by John Michell referred to in the foregoing pages were made by him during his residence at Cambridge ; therefore before the spring of the year 1760 when his Earthquake paper was presented to the Royal Society. He never published any further contributions to geology. It has not unnaturally been inferred that he abandoned that branch of science, in order to devote himself to the severer studies of which the fruits were given in his subsequent papers. But the truth is that his interest in geological questions remained unabated to the end. Proof of this continued zeal is to be found in a long letter of the year 1788, hitherto unpublished, written by Michell to Henry Cavendish, which has fortunately been preserved among the papers of that great philosopher. It vividly indicates how keenly its writer, in his journeys to and from London, kept himself on the watch for any fresh pit, quarry or other exposure of the rocks below the surface.

It appears that Cavendish, for some

years between 1783 and 1793, was in the habit of making excursions with Dr Charles Blagden, for the purpose of tracing the succession and distribution of the strata that underlie the southern counties of England. Of these "Journeys," as he called them, he made notes, which have been preserved. The Cavendish MSS., at the instance of the Cambridge University Press, have been placed by the Duke of Devonshire in the hands of Sir T. Edward Thorpe, with a view to the publication of the chemical papers which they include. The collection of manuscripts contains the letter from Michell, together with the draft of Cavendish's reply. This correspondence, which throws an interesting light on the condition of geological science in this country during the last quarter of the eighteenth century, is now published for the first time[1].

[1] Dr (afterwards Sir) Charles Blagden, who became one of the Secretaries of the Royal Society in

John Michell to the Hon. Henry Cavendish

[14TH AUGUST, 1788]

Dear Sir

Some observations, as I returned from London, having occur'd to me with regard to the Northamptonshire, Lincolnshire, &c. yellow limestone (viz Dr Blagden's, not my yellow limestone) I take the liberty of communicating them to you, though perhaps hardly worth your attention. I could, indeed, have wished, I had been able to give them you with more precision. I lodged one night, in my road, at the Royal Oak, a new house built on Greetham Common,

1784, appears to have acted as assistant or secretarial friend to Cavendish, who settled an annuity upon him and left him a handsome legacy. The MS. of the *Journeys* contains the joint observations made by the two coadjutors. It will be seen from Michell's letter that he was able to set them right on at least one important point, and that Cavendish acknowledged the correction in his reply. As already mentioned, Michell's letter has been alluded to in Wilson's *Life of Cavendish*, pp. 129, 177.

about 7 or 8 years ago, 96 miles from London, which is in the midst of that set of strata, which constitute the Yellow Limestone[1]. When walking in the garden there, I unexpectedly found it to be upon clay[2], and enquiring of the master of the house about it, I found that he had been obliged to sink a ditch, between three and four feet deep, at one side of his garden, as well as to make two or three drains of about the same depth to carry the water into it, in order to prevent it from being so swampy as to be unfit for that purpose; and the water at that time stood some inches deep in some parts of the ditch, though it was in the most droughty part of that time when every-thing about London was so much burnt up, which was also the case in a good

[1] [Apparently a name for the limestones of the Lower Oolite group.]

[2] [There can be little doubt that this was a portion of the "chalky boulder-clay" of the district, lying unconformably upon the various Jurassic rocks, and enclosing flints, bits of coal and many other materials from northern sources.]

measure, though not quite so much so, about Greetham. I the less expected to find things in this state, the land hereabout not being low, and having a moderate declivity, sufficient, I should have thought, if it had not been retain'd by the clayeyness of the soil, to have carried off the water even of a wet season.

I also observed lying about two or three small heaps of pebbles, among which were some flints ; and enquiring of the master of the house, whence they came, he informed me that they were pick'd up from the plough'd fields, which consisted of the same clay with the garden: they were lodged, as I understood, amongst the clay, being found here and there in digging into it. It was not until after you and Dr Blagden mention'd your having seen some specimens of chert, at some place on the coast, I think, amongst this set of strata, that I was aware that any flints were ever found belonging to them, and the flints I met with at

Greetham Common, must, I suppose, be of the same kind with those you consider'd as chert, though I should rather consider them as flints; for though they are opake and had nothing of that horny look, when broken, that the flints from the chalky countries have, yet they have more of the glassy texture, and want that appearance of toughness, which the cherts in general have, so that I should not hesitate to call them flints rather than cherts. At the same time, I can easily conceive that our ideas of them may not so far coincide, but that you might well enough look on them as belonging to the cherts. I however met with, amongst the rest, two or three flints that everybody must look on as such, being, when broken, black and horny, and as perfect as the most perfect of the chalk country flints; they were also roundish like those, and were cover'd with a dark brown coat; whereas the others had no coat, nor any appearance of ever having had one, that I could see, being rather angular and somewhat irregularly shaped.

My landlord also told me, he had been informed (for he had only kept the Inn a year or two himself) that, when the house was built, they had sunk a Well nine yards deep through this bed of clay, before they came to the stone ; the clay may therefore, when compleat, very possibly have been of still greater thickness, but I had no opportunity of learning any further particulars about it. My Landlord also informed me, that he had been told that in sinking the above Well, they had met with in the clay a few small stragling bits of coal, but nothing, as far as he could make out, from the vague account he had been able to procure, and which came through three or four hands, that seem'd to have any tendency towards a regular stratum. This story, however, seems to have induced the owner of the estate (Lord Winchelsea, I think), to try for coal somewhere thereabouts ; for he had had people to bore in search of it, and they had gone to the depth of 130 yards without any success, as I could easily conceive.

This clay did not seem to compose a very uniform stratum, not only consisting of harder and softer parts, but having likewise those flints and pebbles scatter'd through it, in such manner, if I conceived rightly of the matter, as to shew that though they might perhaps have been formed in it originally, yet supposing this to be the case, they must, however, have been somewhat disturbed from their places after their formation, though I neither saw nor could learn circumstances sufficient to form any probable guess concerning the way in which these flints, as well as the other pebbles, which seem'd to contain sand and iron in their composition, were formed. May I not however, consider the circumstances and company in which they are found as rather tending to strengthen my conjecture concerning the origin of flints in general[1] ?

Besides this bed of clay, of the existence of which I was not aware before my last return from town, there is another pretty considerable bed of clay (for I think it is not the same appearing again at another place) which I have often taken notice of, that shows itself in the side of the hill immediately descending towards Grantham, on the east side of it[2]. What is the thickness of this bed I don't know, but from what I have been able to learn concerning it, I should suppose it is not less than the other. There are also found in it, in one part of the stratum, some *Cornua ammonis*, and in another part some selenites ; but these last I pay no great regard to, as they are frequently of a very

[1] [It would have been interesting to know what this conjecture was. How great would have been Michell's astonishment could it then have been revealed to him what is now known about the history of the Boulder-clay which he here describes so minutely.]

[2] [Probably one of the clays of the Upper Lias.]

modern origin, being commonly found in clay, where some vitriolic water oozes or trickles out, provided there is a little calcareous matter likewise for it to unite with. There are a great many Bricks and Tiles made out of this clay for the use of the town of Grantham ; and I imagine, what might otherwise be very well, I think, supposed to be the case, that it is not an accidental mass of clay in that place only[1], but part of a stratum of some extent, for I observed some other Brick-kilns, at a mile or two distance, on the side of a hill, at about the same level. Whether there may not be still more beds of clay in some other parts of this set of strata, I don't know, though from these instances and general analogy, it is not very unlikely there should. Almost immediately to the westward on this side Grantham[2], we again have clay, which is

[1] [Like the tract of Boulder-clay above described.]

[2] [The rock formations to the west of Grantham

continued to the top of Gunnerby Hill, but which, however, must no doubt consist in great part of some kind of stone; for it could not otherwise rise so much as it does in so short a space, viz 70 or 80 yards perpendicular, I apprehend, in the distance of a little more than a mile. There is likewise another set of strata which form another ridge of lower hills, three or four miles nearer this way, about Foston; all these probably contain several beds of clay and under these are found the Lyas, which consists of a great many alternate beds of clay and blue limestone.

I believe I have formerly mention'd it to yourself and Dr Blagden, but not recollecting whether I have before insisted so much upon it, as I might have done, I shall take this opportunity, which the country I have just been mentioning suggests, of observing, that to the westward of all that edge of Dr Blagden's yellow limestone, next our side of the

consist of what are now known as the Upper, Middle and Lower Lias.]

sets of strata which run from north to south through the island of Great Britain, as far as I am acquainted with them, lies the Lyas at no very great distance; though, indeed, with two or three sets of strata, viz those of Gunnerby and Foston, between them; these run into Leicestershire to the south, and to where the Trent falls into the Humber, and the upper part of the Humber to the north, the Lyas being the lowest of all these sets of strata, and all of them lying below the yellow limestone[1] in order, but nowhere having any coal near them; whereas our yellow limestone[2] has no Lyas anywhere under it or near it to the westward of it, but on the contrary, everywhere coal very near the western edge of it, all the way from Leicestershire by the edge of Nottinghamshire and Derbyshire, and a long way into York-

[1] [Now known as the group of Lower Oolites.]

[2] [That is, the Magnesian Limestone of the Permian system which stretches as a broad band from near Nottingham to the mouth of the Tyne, a distance of 150 miles.]

shire, and how much further I don't for certain know ; and in many places, if not everywhere, the coal is found under our yellow limestone, through which they sink in many places in order to come at it.

Since I began to write this letter I received from Mr Beatson of Rotheram, a parcel of the substance he was mentioning to you. He sent, by the person who brought it to me, an apology for not having sent it before, and saying at the same time that it was not yet so good a specimen as he had wished to have sent. As it was directed to me, though it ought perhaps rather to be consider'd as your property, I have taken the liberty of reserving the half of it for myself, which, however, if you want any more than I have sent you, either to make experiments upon, or for any other purpose, I will send you whenever you please. It seems to be in general a good deal harder than the black lead used for pencils, though some of the thin flakes seem to mark pretty well ; probably the difference may be owing to too large a quantity of Iron contained in it; for it appears by it's applying so very strongly as it does to the magnet, to contain a great proportion of that metal.

With best respects to yourself and due comp[s] to all friends when you see them, particularly those of the "Crown and Anchor" and "Cat and Bagpipes" clubs[1], I am, Dear Sir

Your obed[t] humble servant[2]

J. MICHELL

THORNHILL, 14 *Aug[t]* 1788

[1] The Crown and Anchor Tavern, Strand, as already mentioned, was the meeting place of the Royal Society Club at this time and continued to be so for sixty-eight years, from 1780 to 1848. Reference has been made (p. 20) to Michell's frequent appearance at the Club, where he constantly met Cavendish. The "Cat and Bagpipes" was "a public house of considerable notoriety, with this sign. It existed at the corner of Downing Street, next to King's Street. It was also used as a chop-house, and frequented by many of those connected with the public offices in the neighbourhood" (*Notes and Queries*, Nov. 9, 1850, p. 397). But nothing seems to be known of the Club to which Michell refers as meeting there.

[2] This deferential expression, so characteristic

Along with this letter there has been preserved among the Cavendish papers the rough draft of the reply to it sent by the philosopher, which is chiefly interesting as an example of the detailed examination which Henry Cavendish continued for some years to bestow upon the sequence and distribution of the geological formations of the southern half of England. With only lithological characters as a guide, he could hardly fail to make mistakes in the order of superposition.

"I am obliged to you and Mr Beatson for the plumbago and to you for your letter.

I have got some which I received from Wales, part of which, I think, is purer than Mr Beatson's. But the rest consists of flakes of a more sparkling nature than Beatson's and less disposed to mark paper. I have also some which I received under

of the period, was the usual manner in which Michell ended his letters to Cavendish. It is found in the original of his paper of 26th May 1783, which is printed in *Phil. Trans.* vol. 74 (1784), p. 35.

the name of sulphur-iron, and which is much the same to appearance as the latter part of the Welsh specimen. I analysed this and found it to contain more siliceous earth than plumbago, besides a good deal of iron, not so much in the state of plumbago but what it would dissolve in acids.

I suppose it must be the yellow limestone about Bridport in which Dr Bl. told you we found chert. How far it deserves that name I can not say, but to the best of my remembrance it was of a much coarser grain, and had not at all the appearance of flint ; but my memory is too imperfect for me to attempt to describe it to you. As the circumstances relating to it are rather remarkable, I will mention what we saw of it last year.

On descending the chalk hills between Dorchester and Bridport, by the time we got about ½ way to the bottom, we came to the yellow limestone, which seemed separated from the Chalk only by a stratum of clay of no great thickness.

A few miles farther, the stone, though to appearance much the same, was found to be of a siliceous nature, with very little calcareous matter in it. At Lyme the cliffs are blue clay and blue Lyas ; but the top of the hill, which we pass over immediately before we come to Lyme, consists of gravel composed of this chert ; and about a mile to the west of Lyme was a hill with a steep bank towards the sea, the foot of which was blue Lyas with yellow limestone over it, mixed with veins of this chert, so much like limestone that one could hardly distinguish them by the eye ; but it must be observed that this, as well as most of the limestone we saw, is of a hard compact and rather brown kind.

From hence to Sidmouth the soil consisted chiefly of this cherty gravel, but the cliffs on each side of Sidmouth consisted of red rock (the sandy kind consisting of thick strata) ; only on the east side they were covered with a great thickness of the same chert-gravel as the hill by Lyme. From hence we had red rock and red soil, without any chert-gravel to Halldown [Haldon], which is a hill extending from a little to the west of Exeter to near Teignmouth. The upper part of this hill consisted of the above-mentioned chert-gravel, so that it appears that the limestone of this country is very much mixed with chert, a great deal of which seems to have been reduced to gravel and deposited on strata of older formation, at a great distance from the limestone where it was formed. Besides, Halldown, the top of which is covered with this matter, is, I believe, entirely separated from the rest of the country by a broad tract of the red-rock country. In the cliffs between Minehead and Watchett, I saw the red-rock lying immediately under the blue Lyas.

In digging the tunnel for the canal in Gloucestershire, they have found one or more beds of clay between the strata of yellow limestone, and I believe the Chalk is not free from them. A little to the west of Dunstable considerable springs of water break out on the N.W. side of the Chalk hills, about the level of Dunstable.

I believe you must be right in supposing your yellow limestone to be quite distinct from the other. From what I can learn, I believe the N.W. edge of the other, after running from Gunnerby Hill on the E. side of the Trent, crosses the Humber and runs under the Yorkshire Chalk, and appears again about Castle Howard, and so runs to Scarborough, the Chalk in that place lapping over and extending further west than the limestone[1]."

This letter furnishes an example of the detailed manner in which Cavendish conducted his "Journeys." It contains several

[1] The last page and a half of this draft-letter consists of an account of a journey made by Dr Blagden from Dieppe to Paris, with details of his observations on the geological features of the region through which he passed.

interesting original observations. Of these the most remarkable is that which recognised the important overlap of the Cretaceous series of Yorkshire whereby almost the whole of the underlying Jurassic formations are concealed for a space of some twenty miles—a feature in the geological structure of the country of which the full import was not understood for many years after his time. He evidently accepted Michell's opinion that the yellow limestone which immediately overlies the Coal-measures of Derbyshire and Yorkshire could not be the same as that which overlies the Lias, but must belong to a lower platform in the succession of formations.

In the course of years, with his eyes constantly on the alert for fresh light on geological questions, Michell made many original observations that well deserved to be published, but with characteristic modesty he refrained from putting them in print. At the same time, as in his correspondence with Priestley and with Cavendish, he was ready to communicate them to any enquirer who took an interest in the subject. By a happy accident one of these communications to a friend, was committed by this friend to writing and was published seventeen years after Michell had passed away. In August 1810 there appeared, in the *Philosophical Magazine*, a letter from John Farey, Sen., a well-known geologist of the day, enclosing certain notes made by John Smeaton, the eminent engineer, and endorsed by him as "Mr Michell's account of the south of England strata." Farey states that this account was probably made verbally by Michell to his friend Smeaton, very soon after November 1788, and was taken down by Smeaton "on the cover of a recent letter as being the only piece of paper then at hand; for Mr Smeaton's decease in 1792 shows that it must have been prior to that time." The document was as follows:

	Yards
Chalk	120
Golt	50
Sand, of Bedfordshire	10 or 20
Northampton lime and Portland limes lying in several strata	100
Lyas strata	70 or 100
Sand, of Newark ...... about	30
Red Clay of Tuxford	100
Sherewood Forest, pebbles and gravel	50 unequal
Yery fine white sand	uncertain
Roch Abbey and Brotherton limes	100
Coal strata of Yorkshire[1] ...	—

Farey, in communicating this Table, remarked that the "account of the strata imperfect as it is, shews that Mr Michell was acquainted with the principal features of the south of England strata, at an earlier period than anything that has been published on the subject." He adds as

[1] The Northampton limestone belongs to the Inferior Oolites, and the Portland limestone to the Upper Oolites; they both lie above the Lias as shown in the Table. The Keuper and Bunter divisions of the Trias are here correctly placed between the Lias and the Permian Magnesian Limestone.

specially remarkable that Michell should have correctly applied to the strata between Grantham and Balderton the appellation of Lyas, a term not then known or in use nearer than Gloucestershire or Somersetshire, showing that this sagacious observer "had contemplated the identity of the British strata over wide spaces."

Few men, unless they chance to be experienced field-geologists, can fully appreciate the amount of time, skill and labour which the construction of this Table of Strata required. It must represent the result of the journeys of many years over a large part of the southern half of England. It implies an infinite patience and no little lithological deftness in correlating the petrographical characters of the various strata, with such success as to be able to identify the different members at distant parts of their outcrop. The key furnished by organic remains to the chronological sequence of the formations had not yet been discovered by William Smith, who, born two years after Michell's transference to Thornhill, did not begin to publish his epoch-making discovery until the distinguished Rector had passed away. That the Table given above should be imperfect and in some particulars inaccurate does not derogate from the author's credit and originality. He unquestionably established the succession of the main subdivisions of the English Mesozoic formations, and he did this by laborious determinations of the order of superposition and the identity or close resemblance of mineral characters over a wide region, without any help from palaeontological evidence.

Though students of the history of geological discovery in England have been acquainted with Michell's work and have sometimes expressed their high sense of its value, there is reason to think that the pioneer merit of his contributions to geology has never yet been adequately recognised. Lyell indeed has referred to his "original and philosophical" views on earthquake phenomena and has declared that "some of his observations anticipated in so remarkable a manner the theories established forty years afterwards, that his writings would probably have formed an era in the science, if his researches had been uninterrupted. He held, however, his professorship only eight years[1], when his career was suddenly cut short by preferment to a benefice. From that time he appears to have been engaged in his clerical duties, and to have entirely discontinued his scientific pursuits, exemplifying the working of a system still in force at Oxford and Cambridge, where the chairs of mathematics, natural philosophy, chemistry, botany, astronomy, geology, mineralogy and others, being frequently filled by clergymen, the reward of success disqualifies

[1] His tenure of the office, as already shown, was ess than two years.

them, if they conscientiously discharge their new duties, from further advancing the cause of science, and that, too, at the moment when their labours would naturally bear the richest fruits[1]."

The statement in this quotation that from the time of his entering upon his clerical duties, Michell "appears to have entirely discontinued his scientific pursuits" was doubtless based on the fact that after the appearance of his Earthquake paper he never published any further contribution to geological science. We may well believe that his clerical duties were always conscientiously and zealously discharged. But up till near the close of his life he never ceased to pursue his scientific studies. In regard to his geological proclivities we have seen that so far from abandoning that subject he

[1] Lyell's *Principles of Geology*, Tenth Ed. vol. 1, p. 61. To this testimony should be added that of Fitton, *Phil. Mag.* 1832, 1, p. 268: K. A. von Zittel, *Geschichte der Geologie* (1899), pp. 81, 157. Michell is included in the author's *Founders of Geology*, 1897.

continued to prosecute it with a breadth, originality and success which show him to have been the most accomplished English geologist of his time. He was in no hurry to publish his observations though ever willing to communicate them to his friends, and they have come to light almost by accident since his death. He well deserves to be ranked as one of the founders of Geology in England.

But his scientific activities, extending beyond the geological sphere, ranged far and wide through the physical sciences, and his leisure hours at Thornhill were largely devoted to personal research and experiment in that wide domain. Probably a good deal of his original work was never published, but his papers, which found an appropriate place in the *Philosophical Transactions*, have given him a title to high rank among the natural philosophers of the eighteenth century. To the consideration of this side of his achievement I shall now turn.

II

CONTRIBUTIONS TO PHYSICS

It was in the realm of Physics that the originality and brilliance of John Michell's mind found their widest scope. A living writer has recently said: "In the entire century which elapsed between the death of Newton and the scientific activity of Green, the only natural philosopher of distinction who lived and taught at Cambridge was Michell; and for some reason which, at this distance of time, it is difficult to understand fully, Michell's researches seem to have attracted little or no attention among his collegiate contemporaries and successors, who silently acquiesced when his discoveries were attributed to others, and allowed his name to perish entirely from Cambridge tradition[1]." There can at least be no

[1] *A History of the Theories of Aether and Electricity*, by Professor E. T. Whittaker, F.R.S., 1910, p. 167.

doubt that in his lifetime Michell enjoyed the esteem and respect of the most eminent men of science in his day. His distinction as an investigator was promptly recognised, as we have seen, by his early election into the Royal Society, when Henry Cavendish and other men of note became his friends and correspondents. But it was not until after he left Cambridge that his eminence in natural philosophy was displayed in the successive papers which he communicated to the *Philosophical Transactions*.

The consideration of his researches in physical science may be grouped under the heads of (A) Magnetism, (B) Vision, Light, etc., (C) The Density of the Earth. A separate section will be devoted to his investigations in Astronomy.

A. *Magnetism*

In Michell's first published essay in science—the little volume on Artificial Magnets—two of his prominent characteristics were conspicuously shown, originality and modesty. Though from the title of the book it might be supposed to be merely a new method of producing artificial magnets, it yet contained some fresh researches in magnetism including the discovery of the law of attraction which is "the basis of the mathematical theory of Magnetism[1]." The author believed his method of making artificial magnets to be a contrivance of his own, but he admitted that it might prove to be the same as that of his eminent contemporary Dr Gowin Knight. But of much more importance than the originality of

[1] Whittaker, *History of Theories of Aether and Electricity*, p. 55. The full title-page of Michell's work is as follows: "A Treatise of Artificial Magnets; in which is shewn an easy and expeditious Method of making them, superior to the best natural ones, by J. Michell, B.A. Fellow of Queens' College, Cambridge. Printed by J. Bentham, Printer to the University and sold by W. and J. Mount and T. Page on Tower Hill &c. MDCCL. (Price 1/-)." The copy of the volume in the Library of the Royal Society has a MS. note at the foot of the title-page: "Presented March 22, 1750."

his invention was the light which he was able to throw on the laws of magnetism. Thus he found that according to his observations "the magnetical attraction and repulsion are exactly equal to each other." He made and announced the discovery that "the attraction and repulsion of magnets decreases as the squares of the distances from the respective poles increase." Yet he modestly remarks that, although his own experiments made the conclusion very probable, "I do not pretend to lay it down as certain, not having made experiments enough yet to determine it with sufficient exactness[1]." It will be remembered that his contribution to magnetism was one of the grounds set forth in the certificate for his election into the Royal Society, and that one of his sponsors was Dr Gowin Knight, the most noted authority of the day on this branch of science.

[1] *Treatise of Artificial Magnets*, p. 19.

B. *Vision, Light, and Colours*

Interesting proof of the range of Michell's studies in natural philosophy and of the singularly large-minded generosity with which he freely communicated to other fellow-workers the results of his own unpublished researches is furnished by the record of his association with Joseph Priestley. That illustrious philosopher became minister of Mill Hill Chapel, Leeds, in 1767, the same year that saw John Michell settled in the rectory of Thornhill. He had already made known his growing heterodoxy, but he had also shown such striking powers in scientific discussion, particularly in regard to electricity, that the Royal Society had already in 1766 elected him one of its body. Leeds and Thornhill being only a few miles apart, it was natural that the two men of science should become acquainted with each other. In these days it said much for the broad-mindedness of the Rector of Thornhill that he entered into the friendliest relations with the unitarian dissenter. He could hardly fail to be interested in the publication of Priestley's volume on *The History and Present State of Electricity* which made its appearance in this same year 1767. During the lapse of a few years much friendly personal intercourse, as well as correspondence, arose between the two men. Priestley had then begun to collect material for another work on physical science which was published in 1772[1]. During the five years over which the writing of this treatise extended, he frequently consulted the Rector on the various questions which he had to discuss, and he fully acknowledged the value of the assistance which was always willingly forthcoming from that source. He has stated that "in writing the *History of the*

[1] The title of this work is *History and Present State of Discoveries relating to Vision, Light and Colours*, 2 vols. 4to, 1772.

Discoveries relating to Vision, I was much assisted by Mr Michell, the discoverer of the method of making artificial magnets. Living at Thornhill not very far from Leeds, I frequently visited him, and was very happy in his society[1]."

In the two quarto volumes to which Priestley refers in this quotation, he acknowledges in detail his indebtedness to Michell. From his statements and his quotations from the Rector's letters we learn what were Michell's views on a number of physical questions on which he does not appear ever to have himself published anything. Thus in reference to the seat of vision, Priestley remarks: "I shall beg leave to present to my readers some other arguments which escaped the notice of previous observers, but which were suggested to me by my friend Mr Michell[2]." Later in the same

[1] *Life and Correspondence of Joseph Priestley*, by J. T. Rutt, 1831, vol. 1, p. 78.

[2] *Op. cit.* vol. 1, p. 198. Michell as a staunch

volume he states: "My objections to Newton's manner of accounting for the colours of thin plates are of long standing, but the hint of accounting for them in the manner that I have attempted to do [by the doctrine of *attractions* and *repulsions*] was first suggested to me by Mr Michell, agreeably to whose conjectures relating to this subject, I have given the preceding account of the probable cause of these appearances[1]."

Priestley likewise refers to Michell's skill in devising apparatus for the purpose of illustrating or solving physical problems. Thus with regard to another phenomenon of light he states: "Mr Michell some years ago endeavoured to ascertain the momentum of light in a much more accurate manner than those in which M. Homberg and M. Mairan had attempted it; and though his appa-

follower of Newton believed in the corpuscular theory of light.

[1] Vol. I, p. 311.

ratus was disordered by the experiment, and on other accounts, he did not pursue it so far as he had intended, it was not wholly without success ; and the conclusions that may be drawn from it are curious and important[1]."

After describing the apparatus which had been employed, Priestley proceeds to show that the conclusions which its contriver was disposed to draw from his observations, as far as they had gone, pointed to the "mutual penetrability of matter." He states that the ingenious hypothesis of Boscovich on this subject, "or at least one that is the same in everything essential, occurred also to my friend Mr Michell, in a very early period of his life, without his having had any communication with M. Boscovich, or even knowing that there was such a person. These two philosophers had even hit upon the same instances, to confirm and illustrate their hypotheses,

[1] *Op. cit.* p. 387.

especially those relating to contact, light and colours.

"This scheme of the *immateriality of matter*, as it may be called, or rather, the *mutual penetration of matter*, first occurred to Mr Michell on reading Baxter *On the Immateriality of the Soul.* He found that this author's idea of matter was, that it consisted, as it were, of bricks, cemented together by an immaterial mortar. These bricks, if he would be consistent to his own reasoning, were again composed of less bricks, cemented, likewise, by an immaterial mortar and so on *ad infinitum.* This putting Mr Michell upon the consideration of the several appearances of nature, he began to perceive that the bricks were so covered with this immaterial mortar, that if they had any existence at all, it could not possibly be perceived, every *effect* being produced, at least in nine instances in ten certainly, and probably in the tenth also, by this immaterial, spiritual and penetrable mortar.

Instead, therefore, of placing the world upon the giant, the giant upon the tortoise, and the tortoise upon he could not tell what, he placed the world at once upon itself; and finding it still necessary, in order to solve the appearances of nature, to admit of extended and penetrable immaterial substance, if he maintained the impenetrability of matter; and observing farther, that all we perceive by contact, etc. is this penetrable immaterial substance, and not the impenetrable one, he began to think he might as well admit of *penetrable material*, as well as *penetrable immaterial substance*, especially as we know nothing more of the nature of *substance*, than that it is something which supports *properties*, which properties may be whatever we please, provided they be not inconsistent with each other, that is, do not imply the absence of each other. This by no means seemed to be the case in supposing two substances to be in the same place at the same time, without excluding each other; the objection to which is only derived from the resistance we meet with to the touch, and is a prejudice that has taken its rise from that circumstance, and is not unlike the prejudice against the *Antipodes*, derived from the constant experience of bodies falling, as we account it, downwards[1]."

In connection with other problems in light and vision Priestley refers to information supplied to him by Michell and quotes from some of the philosopher's published astronomical papers where these problems are considered.

[1] *Op. cit.* pp. 392–3. As Professor Whittaker has pointed out, Faraday's suggestion that "an ultimate atom may be nothing else than a field of force—electric, magnetic and gravitational—surrounding a point-centre, is substantially the view of Michell and Boscovich." *History of the Theories of Aether and Electricity* (1910), p. 217.

C. *The Density of the Earth*

The most ingenious and most important piece of apparatus devised by John Michell at his Yorkshire home was his bold and original invention of the Torsion-balance with which he proposed to determine the mean density of the Earth. It was probably his last feat in mechanical contrivance, at least he did not live to put it into use. After his death the apparatus passed into the hands of Henry Cavendish who, making some modifications and improvements in it, carried out Michell's purpose with brilliant success, in what has since been known as the "Cavendish experiment." In communicating his account of the experiment to the Royal Society[1], Cavendish, who seemed so indifferent to the recognition of his own scientific work, took care to bear his testimony to the

[1] *Phil. Trans.* vol. 88 (1798), p. 469. The paper was read to the Society on 21st June 1798.

originality of his deceased friend. "Many years ago," so he wrote, "the Rev. John Michell of this Society contrived a method of determining the density of the earth, by rendering sensible the attraction of small quantities of matter; but as he was engaged in other pursuits, he did not complete the apparatus till a short time before his death, and did not live to make any experiments with it. After his death the apparatus came to the Rev. Francis John Hyde Wollaston, Jacksonian Professor at Cambridge, who not having conveniences for making experiments with it, in the manner he could wish, was so good as to give it to me."

"The apparatus is very simple: it consists of a wooden arm, 6 feet long, so as to unite great strength with little weight. This arm is suspended in an horizontal position, by a slender wire, 40 inches long, and to each extremity is hung a leaden ball, about 2 inches in diameter, and the whole is enclosed in a little wooden case to defend it from the wind.

"As no more force is required to make this arm turn round on its centre than what is necessary to twist the suspending wire, it is plain that if the wire is sufficiently slender, the most minute force, such as the attraction of a leaden weight a few inches in diameter, will be sufficient to draw the arm sensibly aside. The weights which Mr Michell intended to use were 8 inches in diameter. One of these was to be placed on one side of the case opposite to one of the balls, and as near it as could conveniently be done, and the other on the other side, opposite to the other ball, so that the attraction of both these weights would conspire in drawing the arm aside; and, when its position, as affected by these weights, was ascertained, the weights were to be removed to the other side of the case, so as to draw the arm the contrary way, and the position of the arm was to be again determined ; and consequently, half the difference of these positions would shew how much the arm was drawn aside by the attraction of the weights.

"In order to determine from hence the density of the Earth, it is necessary to ascertain what force is required to draw the arm aside through a given space. This Mr Michell intended to do, by putting the arm in motion, and observing the time of its vibrations, from which it may easily be computed[1]."

"Mr Michell had prepared two wooden stands on which the leaden weights were to be supported, and pushed forwards, till they came almost in contact with the case; but he seems to have intended to move them by the hand.

"As the force with which the balls

[1] "Mr Coulomb has, in a variety of cases, used a contrivance of this kind for trying small attractions; but Mr Michell informed me of his intention of making this experiment, and of the method he intended to use, before the publication of any of Mr Coulomb's experiments." [Note by Cavendish.]

are attracted by these weights is excessively minute, not more than $\frac{1}{50,000,000}$ of their weight, it is plain that a very minute disturbing force will be sufficient to destroy the success of the experiment; and from the following experiments it will appear, that the disturbing force most difficult to guard against is that arising from the variations of heat and cold; for, if one side of the case is warmer than the other, the air in contact with it will be rarefied, and, in consequence, will ascend, while that on the other side will descend, and produce a current which will draw the arm sensibly aside.

"As I was convinced of the necessity of guarding against this source of error, I resolved to place the apparatus in a room which should remain constantly shut, and to observe the motion of the arm from without, by means of a telescope; and to suspend the leaden weights in such a manner, that I could move them without entering the room. This difference in the manner of observing, rendered it necessary to make some alteration in Mr Michell's apparatus; and as there were some parts of it which I thought not so convenient as could be wished, I chose to make the greatest part of it afresh."

The "Cavendish experiment" has become famous in the annals of physical science. One of the most appreciative accounts of it and of Michell's share in preparing for it was penned by the distinguished Professor of Natural Philosophy in the University of Edinburgh, James David Forbes, more than half a century after both Michell and Cavendish had been laid in the grave. The concluding sentences of his narrative may be quoted here: "Cavendish conducted the experiment with his usual patience, judgment and success; he found the joint attraction of the small balls and large spheres to be about $\frac{1}{4300}$ of a grain, their centres being 8·85 inches apart, and he thence computed the density of the Earth to be 5·48 times that of water. Cavendish's paper is, as usual, a model of precision, lucidity and conciseness. It would be difficult to mention in the whole range of physics a more beautiful and more important experiment[1]."

Since Cavendish improved Michell's apparatus and first put it to the use for which its designer constructed it, the experiment has been repeated by several observers with an approximately similar result[2]. The most recent repetition is that of Mr C. V. Boys. By an ingenious reconstruction of apparatus and availing himself of the great sensibility obtained by the use of quartz-fibres instead of metal wires this accomplished physicist

[1] Professor Forbes' description is contained in the Sixth Dissertation of the Eighth Edition of the *Encyclopaedia Britannica*, p. 834.

[2] See Reich, *Compt. rendus*, 1837, p. 697; Baily, *Mem. Astron. Soc.* vol. XIV; *Phil. Mag.* XXI (1842), p. 111; Cornu, *Compt. rend.* vol. 86, pp. 571, 699, 1001.

has computed the mean density of the Earth to be 5·5270[1].

III

CONTRIBUTIONS TO ASTRONOMY

THE studies pursued by John Michell in this branch of science were marked by his characteristic originality and insight. Not only was he an actual observer of the heavens, working with a reflecting telescope of his own construction, but in his theoretical discussion of stellar phenomena he introduced the mathematical computation of probabilities, and showed sometimes a remarkable prescience that seems to anticipate the discoveries of more recent times. Reference has already been made to the family tradition that Michell gave William Herschel his first lessons in Astronomy and taught him the art of making reflectors. Before entering

[1] Boys, "On the Newtonian Constant of Gravitation," *Phil. Trans.* vol. 186 (1896); see also *Proc. Roy. Soc.* vol. 46, p. 253; *Proc. Roy. Instit.* vol. XIV (1894).

on the consideration of Michell's own astronomical work it may be convenient if we take note of what were the actual personal relations of these two astronomers.

It is now clearly established that they started quite independently of each other in the actual construction and employment of the reflecting telescope. We do not know when and under what conditions the Rector of Thornhill began to construct the large instrument which ultimately became the property of Herschel, but it would appear that he had made considerable progress, if he had not completed it before 1781. Herschel did not begin to study astronomy until 1773 when he was still actively engaged in the multifarious duties of his musical profession at Bath. In the following year he began to grind specula[1]. After six years, during which he worked laboriously with his telescope, he was able

[1] *The Scientific Papers of Sir William Herschel*, vol. I, pp. xxxi–xxxii.

in the summer of 1780 to send to the Royal Society two papers in which the results of his first researches were given[1]. These papers revealed to the world the advent of a new astronomer of unusual promise. They would probably be known and appreciated by Michell, for they appeared in the *Philosophical Transactions*. But, as already stated (p. 18), they were more pointedly brought to his notice by his friend Dr Watson, who had interested himself at Bath in the work of the precocious astronomical musician. Herschel took advantage of the opening provided by Michell's letter to him of 21st January 1781. To two of his letters Michell sent him a long reply (12th April) dealing with the construction of mirrors and the relative merits of different types for large and small apertures. Only a month before this letter was written Herschel had made his great discovery of Uranus, and had thus leaped into a foremost place among

[1] *Phil. Trans.* vols. 70 and 71.

the astronomers of the world. The correspondence between him and the Rector of Thornhill does not appear to have been maintained; but as Herschel was elected a Fellow of the Royal Society on 6th December 1781, the two men of science would now have opportunities of personal intercourse at the meetings of the Society and the convivial gatherings of the Royal Society Club. Herschel in subsequent years took occasion, in a paper read before the Royal Society, to refer appreciatively to the work done at Thornhill. "Mr Michell," he said, "has also considered the stars as gathered together into groups (*Phil. Trans.* vol. 57, 1767, p. 249) which agrees with the subdivision of our great system here pointed out. He founds an elegant proof of this on the computation of probabilities, and mentions the Pleiades, the Praesepe Cancri, and the nebula (or cluster of stars) in the hilt of Perseus's sword as instances[1]."

[1] *Phil. Trans.* 75 (1785).

The only record which I have been able to recover of an actual meeting of the two astronomers was one made by Herschel during a tour in the year 1792 when he passed through Thornhill and called at the rectory. But the Rector, now near the close of his life, was disabled and frail. Herschel has noted: "We saw Mr Michell's telescope; it is on an equatorial stand, being without cover behind. I put my hand into the opening and felt the face of the object speculum so wet as to moisten my fingers. Mr Michell was very indifferent in health."

When Herschel, in the course of a holiday trip with his wife next year, spent a couple of hours at the place, Michell had already passed away, and the instruments that had gradually been accumulated at the rectory were about to be removed. He took another look at the collection and made a note that he had "bought Mr Michell's great tele-

scope and paid Mr Turton 30 pounds[1]."
It is interesting to know that the instrument was put into good order and was used in his subsequent researches by the great astronomer into whose hands it had come.

We may now pass on to consider Michell's genius for astronomy as displayed in the papers which he communicated to the Royal Society and which duly appeared in the *Philosophical Transactions*. I am glad to be able to present the following estimate of these papers, which at my request has been prepared for this Memoir by my friend Sir Joseph Larmor.

"In designing his apparatus to measure the gravitational attraction of a globe of lead, and thence to deduce the mean density of the Earth, Michell was the pioneer in the standard method of determining very small forces by taking advantage of the torsion produced by them

[1] *Herschel's Scientific Papers*, vol. I, p. lx.

in a wire. It was shortly afterwards, as Cavendish remarks, that Coulomb applied the same principle, in a classical series of experiments, to the exact determination of electric and magnetic attractions: and, in various more convenient forms, it is now one of the main resources of delicate physical measurement. But Michell's (and Cavendish's) mastery of it, and his just anticipation of its power, went far beyond his age; he designed and constructed appliances with confidence, for a precise estimation of forces so minute that they could hardly even be detected in any other way: even nowadays his application of the principle to gravitation demands the resources of a master.

"It is to be expected that a man who could confidently engage in preparations to weigh a ball of lead against one of the celestial bodies would be capable of deep views on other astronomical questions. An examination of his Memoir of 1767 confirms this surmise[1]. As regards general astronomical speculation on stellar systems and their nature, it gives him a place alongside Huygens, Wright and Kant[2]. Further, in more definite fields, it credits him with initiation of the application of mathematical methods, resting on probability and statistics, to the celestial systems. The quantity of material which had then been accumulated was far too small for wide statistical inferences of much certainty; yet Michell amply demonstrated, for the first time[3], the

[1] The title of this paper is as follows: "An inquiry into the probable Parallax and Magnitude of the Fixed Stars from the quantity of Light which they afford us, and the particular circumstances of their situation." *Phil. Trans.* vol. 57 (1767), p. 234. Herschel's reference to this paper has been referred to *ante*, p. 95. Later references will be found in Todhunter's *History of the Mathematical Theory of Probability* (1865), where it is stated that the paper had "attracted considerable attention." Michell's method of enquiry is there quoted and his results are given (pp. 332, 393, 491).

[2] See R. Grant, *History of Physical Astronomy*, pp. 543, 558, 559.

[3] Grant, *op. cit.* p. 547.

most fundamental fact of stellar cosmogony, the existence of physically-connected stellar groups. In the case of the conspicuous pairs of adjacent stars (the so-called double stars) he anticipated that orbital revolution round each other, owing to their mutual gravitation, would in time be detected,—a prediction afterwards brilliantly realised on a grand scale by Sir William Herschel. He even pointed out that knowledge of the period of their orbital revolution, combined with their distance from the solar system, would provide means of determining the mass of such a stellar pair in comparison with the mass of the Sun[1],—a problem which is being worked out into exact knowledge by aid of refined determinations of parallax in our own time.

"These considerations occur in the course of discussion of a plan for estimating the distances of the stars by com-

[1] *Phil. Trans.* 1784, p. 36, *et seq.*

paring their brightness with that of the Sun, on the assumption that they give out an amount of light not greatly different from his. This method had, it seems[1], been first suggested by James Gregory; it was applied roughly by Huygens to Sirius; and it attracted the attention of Lambert and later of Olbers, as well as that of Michell. In Michell's argument the planet Saturn, whose size and distance are known from the Newtonian theory, and whose brightness relative to the Sun could thus be estimated, was used as an intermediary; for it would be impossible to compare directly the dazzling brightness of the Sun with the amount of light received from a star. These astronomers all agree in assigning a parallax less than half a second of arc to the brightest stars; and this is in fact near the values that are now known for the very few nearest stars, which are thus at a distance from our system of about a million times that of

[1] Grant, *op. cit.* p. 547.

the Earth from the Sun, while most stars are very many times more remote.

"Michell was the first[1] to propound, in the same Memoir, just views as to the simple proportionality between the faintness of the stars just visible in a telescope and the area of its aperture, no other circumstance being essentially concerned. He initiated the application of this principle to the estimation of the distribution of the stars at different distances in the depths of space,—a task afterwards carried out so tenaciously and brilliantly in the 'star-gauging' of Sir William Herschel. He concluded from a discussion of probabilities that the bright stars were more numerous around our system than a uniform distribution in the celestial spaces would permit; and he inferred that most of the bright stars that did not obviously belong to star-groups were our nearer neighbours, and constituted a stellar system of which our own solar system is

[1] Grant, *op. cit.* p. 543.

a part; while the fainter stars in the depths of space may be grouped in other stellar systems. Thus he thought the nebulae were separated universes of stars, so far away as to defy resolution into their components. Modern astronomical theories are now moving, of course far more definitely, along similar lines, fortified by the immense masses of facts relating to distances, motions and constitutions of the stars and nebulae, which are provided by the photographic plate and the spectroscope in conjunction with large telescopes.

"In Michell's day the available data were utterly inadequate to guide to safe statistical conclusions on matters of such delicate inference. But the mathematical modes of reasoning in his Memoir of 1767 are still of much interest in the light of modern knowledge, especially as they are illustrated by a discussion of the group of the Pleiades, as it is presented to the naked eye and also in telescopes of various apertures. It may be claimed that these modes of reasoning give Michell a place as the early pioneer in the great modern problem of the configuration and structure of the universe, which first rose to prominence twenty years afterwards, by the labours of Sir William Herschel, founded on similar views.

"In regard to optics, Michell was a thoroughgoing Newtonian, as was natural in his time. Light for him consisted of corpuscles projected from the luminous body, rather than waves propagated through an aether. He even thought that, like everything material, they must be subject to gravitation; and he developed a speculation that the velocities of the corpuscles shot out from one of the larger stars must be sensibly diminished by the backward pull of its attraction, and thus be more deviated by a glass prism, a supposition which he proposed to test. At the end of his Memoir of 1767 (p. 261) he even suggests that the 'twinkling of the fixed stars' is due to the small number of luminous corpuscles received by the eye which might be only a few per second. These corpuscular optical speculations now carry special interest as a curiously definite foreshadowing of the work on electric radio-activity, in which Thomson, Rutherford and others have actually controlled the velocities of the electric corpuscles by the agency of field of force, and have directly counted the numbers of them that are shot out from active matter.

"In his wide outlook over the field of nature, in the extent of knowledge that was linked together in his active interests, Michell was a true disciple of the British school of physical science, the contemporary members of which were largely his personal friends. They were falling behind in mathematical analysis, owing to too conservative partiality for the geometrical methods of their master Newton. While the great analysts of the continent were closely engaged in the expansion of the infinitesimal calculus and its improvement by application to the verification and prediction of the motions of the solar system, the mathematicians of Britain had time for wider, though less intricate contemplation of the correlations of natural phenomena, not seldom leading into general views which subsequent times were to develop with fuller knowledge."

INDEX

Astronomy, Michell's contributions to, 91–105
Balderton, 67
Baxter, on the immateriality of the Soul, 81
Birch, Dr T., 11
Blagden, Sir Charles, 46, 49, 55, 60, 63
Boscovich, R. G., 80, 83
Boys, C. V., 90
Bridport, 60
Brotherton, 66
Castle Howard, 63
Cavendish, Hon. Henry, 3, 12, 20, 45, 47, 59, 84–90
Chalk, 60, 63, 64, 66
Coulomb, C. A. de, 87, 97
Dorchester, 60
Dunstable, 63
Earth, determination of mean density of, 84
Earthquakes, Michell on, 25–39
Exeter, 62
Faraday, M., 83
Farey, John, 65, 66
Fitton, Dr, 70
Fitzpatrick, Rev. T. C., 4
Forbes, James David, 89
Foston, 55, 56
Gault, 66
Geology, Michell's contributions to, 24–58
Grantham, 54, 67
Greetham Common, 50

Gunnerby Hill, 55, 56, 63
Hadley, Prof. J., 11
Haldon Hill, 62
Herschel, Sir William, 14, 19, 91–96, 101, 103
Homberg, M., 79
Humber, 63
Knight, Dr Gowin, 11, 74
Larmor, Sir Joseph, 96
Leeds, 76
Lias, 53, 61, 62, 66, 67
Light, Michell's conception of, 79, 103
Lyell, Sir Charles, 69
Lyme, 61
Magnetism, Michell on, 73
Mairan, M., 79
Maty, Dr M., 11
Michell, John, earliest account of, 3; at Queens' College, Cambridge, 4–6; Woodwardian Professor of Geology, 6; his personal appearance, 7; Fellow of Queens', 8; his treatise on Magnets, 8, 73; his Essay on Earthquakes, 9, 25; elected into the Royal Society, 11, 73; Rector of Compton, 12; his Marriage, 12; Rector of Havant, 13; his daughter, 13, 14; family traditions, 13; Rector of Thornhill, 13; his connection with London, 20; his early interest in geology, 21; his death and epitaph at Thornhill, 23

His contributions to Geology, 24–58; on the causes and phenomena of Earthquakes, 25; on the structure of the stratified part of the Earth's crust, 39; his contributions to Physics, 72–90; on Magnetism, 73; on Vision, Light, etc., 76; on the immateriality of Matter, 81; his Torsion-balance for determining the mean density of the Earth, 84; his contributions to Astronomy, 91–105

Minehead, 62
Physics, Michell's contributions to, 72–90
Priestley, Joseph, 3, 76, 77, 78–83
Queens' College, Cambridge, 3, 14
Roch Abbey, 66
Royal Society, 2, 7, 9, 10, 20, 25, 73, 76, 94
Royal Society Club, 2, 20, 58, 94
Scarborough, 68
Selenite, 53
Sherewood Forest, 66
Sidmouth, 61
Sills or Intrusive Sheets, 37
Smeaton, John, 65
Smith, William, 68
Stars, twinkling of, 103
Strachey, John, 39
Thornhill Rectory, 13, 22, 23, 58, 76, 95
Thorpe, Sir T. Edward, 46
Torsion-balance of Michell, 84, 89, 96
Tuxford, 66
Vapour, subterranean, 29, 36
Vision, Light, etc., Michell on, 76–83
Watchett, 62
Wernerian geognosy, 27
Whin Sill of North England, 38
Whittaker, Prof. E. T., 72, 83
Wilson, George, biographer of Cavendish, 3, 47
Wollaston, F. J. Hyde, 85
Woodward, John, 6, 40
Wray, Dr D., 11
"Yellow Limestone," (Jurassic) 48, 60, (Permian) 56, 63, 64
Zittel, K. A. von, 70

www.ingramcontent.com/pod-product-compliance
Ingram Content Group UK Ltd.
Pitfield, Milton Keynes, MK11 3LW, UK
UKHW040656180125
453697UK00010B/208

9 781107 623781